Gen

MAY 24 2001

FOR YOUR HOME

SMART STORAGE

FOR YOUR HOME

SMART STORAGE

Lisa Skolnik

FRIEDMAN/FAIRFAX
PUBLISHERS

Dedication

To my mother, Dorothy Zuckert, who never throws anything away. I am eternally grateful for both the substance and the sentiment wrapped up in the objects she has kept in storage.

Acknowledgments

Thank you to the talented and diligent staff at Michael Friedman Publishing Group, particularly my editor, Reka Simonsen, for her persistence in obtaining remarkable pictures.

A FRIEDMAN/FAIRFAX BOOK

©1999 by Michael Friedman Publishing Group, Inc.

All rights reserved. No part of this publication may be reproduced, stored in a retrieval system, or transmitted, in any form or by any means, electronic, mechanical, photocopying, recording, or otherwise, without prior written permission from the publisher.

Library of Congress Cataloging-in-Publication Data.

Skolnik, Lisa.
Smart storage / by Lisa Skolnik.
    p. cm. – (For your home)
Includes bibliographical references and index.
ISBN 1-56799-718-X
1. Interior decoration.   2. Storage in the home.   I. Title.   II. Series.
NK2113.S56 1999
648'.8–dc21   99-18453
                                                            CIP

Editor: Reka Simonsen
Art Director: Jeff Batzli
Designer: Meredith Miller
Photography Editor: Jennifer Bove
Production Manager: Camille Lee

Color separations by Fine Art Repro House Co., Ltd.
Printed in Hong Kong by Midas Printing Limited

1 3 5 7 9 10 8 6 4 2

For bulk purchases and special sales, please contact:
Friedman/Fairfax Publishers
Attention: Sales Department
15 West 26th Street
New York, New York 10010
212/685-6610 FAX 212/685-1307

Visit our website:
http://www.metrobooks.com

Table of Contents

Introduction ◼ 6

Kitchens ◼ 10

Living and Dining Areas ◼ 28

Bedrooms and Baths ◼ 44

Special Spaces ◼ 58

Sources ◼ 70

Index ◼ 72

Introduction

Everybody has something to store, from the avowed minimalist to the inveterate collector. It's virtually impossible to go through life without accumulating a few things along the way, such as a small wardrobe or selected decorations for the home, but most of us acquire far more than just the basics. The shelves and bookcases in our rooms are usually filled to the brim with the ordinary assortment of items, and sometimes with more remarkable collections of objects we just can't resist. There are the bare necessities, such as staples for a week's worth of meals, plus the prosaic elements of everyday life: kitchen utensils, toiletries, cleaning supplies, and so on. And lest we forget, in this age of physical fitness and electronic commuting, there are extras to contend with, too. Sporting goods of every ilk and home offices, complete with a whole range of equipment and supplies, create the need for even more storage solutions.

Now, more than ever, it's critical to keep clutter from overwhelming our environments by employing some form of storage. Storage options are many and varied, and several considerations will influence your choices. The architecture and size of your home, the nature and makeup of your family, the requirements of your lifestyle, and your own personal preference will all play a part in the decisions you make.

Different kinds of residences allow for, and incorporate, different kinds of storage. Apartments lack the roomy expanses afforded by homes that have basements and attics, but many newer condominiums and lofts are built with empty spaces just for storage. Older homes often have smaller closets and bathrooms, but they make up for this deficit with larger rooms that can incorporate several kinds of storage. Many also possess nooks and crannies, such as large foyers or stairway landings, that can be used in alternative ways.

Family size is another obvious influence on storage needs. Children always take more space than expected, especially when they're small and have tiny togs but

Opposite: Tansu chests, Japanese storage chests that are made in small, individual sections, are a particularly handsome form of freestanding storage. In this entry hall, several sections have been stacked to create an elegant, single large storage unit. The rich wood and decorative construction of the chests' sliding doors echo the design of the doors at the end of the hallway, which helps unify the space.

bulky gear, such as strollers, swings, playpens, and activity toys. Older children may need space for hoards of their own. Plus, the more souls in a household, the more provisions, clothing, and linens there are to stock in kitchens, closets, and cabinets.

Lifestyles may seem unrelated to storage, but they actually have quite an impact on how a home is used. Such avocations as athletics, gardening, woodworking, reading, painting, sewing, knitting, and surfing the net all have equipment needs. If the gear or supplies these pastimes call for aren't organized within reach, it can be difficult to really enjoy these activities.

Personal preference affects every aspect of storage. Some people favor storage that incorporates display, while others like to hide everything away, but the issue is not quite this simple. The potential display value of everyday objects is often overlooked, yet this is a very useful technique. Not only does it offer easy access to frequently used items, but if properly executed, it can make a space or a collection of belongings look remarkable. However, if too much is left in the open, the space looks cluttered and the objects get dusty and unattractive.

Concealed storage also has its strengths and weaknesses. It can be elegant and sleek, with everything in its own spot, but it often necessitates costly built-ins, and putting things back where they belong requires discipline. There's always the option of a happy medium, where prized possessions are put on show and mundane items are stowed away, but achieving this balance isn't easy.

So how can you get adequate and appropriate storage that meets all of these needs? Start with the basics. Organizing a home doesn't always call for major upheavals or exorbitant expense, but it does call for forethought, planning, and prudence. A little creativity, coupled with a lot of discipline, can help matters. Experts say that the first step is to take stock of everything you own. Go through each area of your home and analyze and purge. Most likely, much of what you own can be pared down, especially since there is no reason to save what you don't like, need, or use.

Next, assess what's left, and separate everything into categories based on how frequently it is used. Items on demand every day need to be readily accessible, while things used on a seasonal basis or sporadically can be put away. It makes good sense to rotate these things in and out of storage twice a year, since this practice will help keep you organized by forcing you to view and evaluate your belongings regularly. Keep a list of what gets stowed in out-of-the-way storage, to make retrieval easier and to keep you from forgetting what you have.

After going through this process, you should be left with what really counts, and it's up to you to make the best of it. The next step is to note how you use your rooms, to get an idea of the space you have at your disposal. Reallocating these spaces sometimes makes sense. A formal dining room can be combined with—or even turned into—a library for someone with a huge book collection. A master suite can become a combined bedroom

Introduction

and a play area for several children. Often, homes have spaces that aren't part of specific rooms but offer up lots of square footage to adapt to storage, such as expansive foyers, areas under stairs, extra-large landings, or spacious circulation corridors. These are prime spots to put to good use, but be careful that you don't hinder the important traffic patterns in your home.

The last step to making the most of your space is to provide plenty of smaller but no less important storage options. Invest in under-the-bed boxes or baskets and over-the-door racks. The areas in each room that house clusters of stuff, such as cabinets, closets, and shelves, must be planned, organized, arranged, and constantly and consistently pruned. After all, we always need more storage space than we actually have, so it's important to keep belongings in check.

Below: THANKS TO ITS SINGULAR APPEARANCE, THIS DISTRESSED METAL CABINET DOES DOUBLE DUTY: IT CREATES PLENTY OF STORAGE SPACE AND GIVES THE ROOM CHARACTER AND STYLE. THE LOW-SLUNG CABINET KEEPS OFFICE PARAPHERNALIA HIDDEN FROM SIGHT, WHILE THE TOP BECOMES AN EXPANSIVE DISPLAY SPACE.

Kitchens

Like the hearth of yesteryear, the kitchen of today is the heart of the home. Every family member uses it at some time during the day, and the functions performed in this space are many and varied. Traditionally, the kitchen has been used for preparing and enjoying meals, but now it encompasses all sorts of pastimes. Depending on its size and scope, parts of the kitchen can act as a home office, craft area, play room, and social center throughout the day. Thus it can be the busiest room in a residence, and therefore must be well planned and carefully organized.

Storage is the key to the success of this room, but even so it comes second to space planning. Before the requisite cabinets, shelves, or cupboards can be integrated into the kitchen, it's critical to define the principal work areas of the space: the sink, cooking apparatus, and refrigerator. Once these are configured, it's possible to locate storage in logical clusters around them.

Most kitchen storage includes built-in cabinetry both above and below the stove, counters, and other work surfaces. The beauty of this system is that these cabinets can both hold all and hide all as needed. If fronted with transparent doors, they can also show all in a decorative fashion. Another plus is that a wide range of interior fittings can be used to customize them, such as drawers, dividers, shelves, racks, and carousels. Such special features as pull-out components and chopping blocks can also be integrated into their design.

Unfitted kitchens, comprised of various types of freestanding storage, are another option, and allow great flexibility for those who like to rearrange their surroundings or who move frequently. Armoires, hutches, and other freestanding pieces can give the kitchen a furnished look rather than a utilitarian "fitted" look, which may be more to your taste. They also offer you a wider choice of

Opposite: Substance and style merge in this eat-in kitchen, which has large, efficiently configured, and attractive work spaces and storage areas. The expansive wraparound granite-topped counter can accommodate several people and a range of tasks. Glass-fronted cabinets display dishes and stemware, while open shelves built into the island showcase a collection of fine china. Drawers and cabinets with solid wooden doors keep the less decorative but necessary kitchen accoutrements hidden away.

decorative styles than built-in cabinetry does, since these pieces can range from contemporary to antique. In fact, the kitchen can become a much more eclectic environment with this approach.

There are many ways to optimize the kitchen's storage potential. Whether you prefer built-in cabinets or freestanding furnishings, the first and most important thing is to make sure that the storage units are well organized and put to the best possible use. Then look at every other nook and cranny. Backs of doors can be fitted with shelves, racks, or grids to store canned or dry foods and cleaning products. Racks suspended from the ceiling can keep pots, pans, and utensils within reach without cluttering up work surfaces. The bottoms of hanging cabinets can be fitted with hooks or holders for wine glasses, coffee cups, spice racks, or utensils. Walls can be fitted with grids or hooks for all sorts of items. Keep in mind that all of these techniques require strong walls and proper installation, or the weight of the equipment will pull the racks or hooks out of the walls.

Left: Industrial shelving can fulfill several objectives. It offers plenty of raw storage space, is cost effective, is extremely flexible since it can be configured to fit any space, and is neutral enough to be integrated into a myriad of decors. Most importantly, it has great visual impact if it's handled creatively. Here, like items are grouped together and in some cases stored in woven baskets to create a visually arresting display. The earthy baskets also offer a nice contrast to the hard-edged steel shelves.

Kitchens

Right: A number of clutter-busting techniques can be employed in the kitchen to make assorted items seem more manageable. Use large stoneware jars to house utensils, butcher-block safety stands to store knives, the tops of cabinets to stash pretty crockery, and interior shelves to accommodate spices and condiments. Attractive utensils can also be hung on a wooden wall to display them without taking up work space.

KITCHENS

Left: THE ISLAND IN THIS AIRY KITCHEN IS FITTED WITH A COOKTOP AND BOTH OPEN AND ENCLOSED STORAGE SPACES. CABINETS UNDERNEATH THE COOKTOP HOLD POTS AND PANS, WHILE OPEN SHELVES HOUSE COOKBOOKS AND DISPLAY FAVORITE PIECES OF POTTERY. THIS KITCHEN HAS PLENTY OF OTHER ATTRACTIVE AND PRACTICAL STORAGE SPACES AS WELL: NOTE THE OPEN SHELF THAT RUNS OVER THE DOORWAY AND THE BASKET-FILLED NICHE ABOVE THE REFRIGERATOR.

Below: IT'S IMPORTANT TO CREATE STORAGE SPACE IN A KITCHEN WHEN NONE IS AVAILABLE, AND IT'S NOT ALWAYS NECESSARY TO RESORT TO ELABORATE, EXPENSIVE, OR TRICKY SOLUTIONS. HERE, A NARROW THOROUGHFARE IS FITTED WITH EQUALLY SLENDER SHELVES THAT ARE SIMPLE, STRAIGHTFORWARD, COST EFFECTIVE, AND EASILY ACCOMMODATE ALL THE NECESSITIES. THE KITCHEN ACCOUTREMENTS HAVE MORE POWER IN A HUGE GROUP THAN ON THEIR OWN; THEY LOOK QUITE ENGAGING INSTALLED EN MASSE.

Kitchens

Left: Center islands come in all shapes and sizes, and can easily be adapted for special needs or specific tasks. This small enclosed island accommodates a sink, additional counter space, and out-of-sight storage. The cozy window seat is an ideal place for a friend or family member to sit and visit with the cook.

KITCHENS

Below: An unconventional storage system can break the rules and still be practical. This prosaic kitchen alcove becomes anything but ordinary, thanks to an unusual setup that combines an angled counter and unique circular shelves suspended from a pole. Now the space boasts a breakfast bar and plenty of storage with undeniable style.

Above: Thanks to all sorts of specialty storage devices, it's possible to mine every nook and cranny in a kitchen. Here, the otherwise dead space of a tight corner is transformed into a storage center and work station, thanks to a built-in cupboard. It is sensibly fitted out with small divided spaces for dishes, as well as a towel rack and a pull-out cutting board.

Kitchens

Above: There are many ways to employ customized wall systems to compensate for small or shallow counters. In this kitchen, an elegant but narrow stainless steel counter is kept clutter-free with two banks of glass and steel shelves set along the wall above it. Matching canisters made of the same materials continue the clean, streamlined feeling. The glass and steel used to create the storage system emphasize the contemporary decor and enhance the light, uncluttered quality of the space.

KITCHENS

Right: Virtually every available inch of wall space in this kitchen is devoted to cabinetry, which contributes to the airy feeling of the room, thanks to the liberal use of glass-faced doors. A patterned floor in rich, burnished colors offers a counterpoint to the cool white of the cabinetry. The blend of these elements creates a sophisticated environment that manages to be warm and refined at the same time.

KITCHENS

Opposite: Simple, understated cabinetry in blond wood blends in nicely with the glossy beige tiles used for the backsplash in this kitchen. Glass doors add variety to the mix and have the visual effect of enlarging the space. They also make good sense, since it's easy to see where everything is. Matching glass canisters store dry foods in an organized and attractive manner.

Above, left: Think creatively. Here, a charming vintage wooden medicine cabinet with oddly shaped compartments is recycled to become a spice chest. Its eccentric configuration is an asset given the nature of the items it now holds.

Above, right: Make the most of every bit of space you have. This tiny nook has been fitted with a small bar sink, a shallow shelf that displays a few accessories, and overhead racks that store a surprising amount of stemware. A cheery green and cream palette, warmed up with dashes of red, makes the space feel cozy rather than cramped.

Kitchens

Above: A small alcove situated between a cooktop and a window becomes a full-service storage center for various foodstuffs, thanks to cleverly conceived shelves that are based on a simple bracket system but customized in a striking style. Because they are mounted on the wall, the shelves can extend over the cooktop to reclaim unused space. Inset colanders and unique drawers create textural interest, and the use of vibrant colors—on the walls and for the decorative glassware—enlivens the small room.

Opposite: A novel storage system can make a routine kitchen remarkable, and is often relatively easy to execute. Here, color and form lend a postmodern slant to a straightforward setup. Several soft tints are used on the built-in cabinetry, which give it the panache of a carefully crafted piece of furniture. Open shelves allow attractive dishes to be on display and add a sculptural component to the mix.

Kitchens

KITCHENS

Opposite: Storage can be used to make a design statement. Here, a custom-built system that alternates rich honey-colored wood with glossy blue enamel for decorative emphasis is used to create peak efficiency in a kitchen. Even the island is configured to make maximum use of the space and provide extra storage.

Below: Use your imagination when the right item of furniture presents itself. A beautifully weathered piece can be mined for additional storage. This vintage shoe rack has been put to use as a wine rack, but it would be equally appropriate for dishes, pots and pans, or cookbooks.

Above: A vintage cupboard gets a design upgrade from the architectural elements and decorative accessories that surround it. The lovely freestanding piece is squeezed between a built-in cupboard and a doorway, and an overhead shelf spans the distance between these two elements to expand the available storage. Baskets enhance the rustic appeal of the cupboard and provide a simple, attractive decoration.

Kitchens

Right: A wall next to a stove is the perfect place to store cooking implements and ingredients where they're needed most. A plain steel shelf with movable hooks keeps utensils handy, while tiered stands allow the cook to find the desired spice easily. The brick wall turns the expanse into a striking backdrop for these everyday items.

Below: The concept of an overhead rack is applied to the wall in this kitchen with remarkable results. The rack blends function with form, as the easily accessible kitchen accoutrements make a visually arresting display. Best of all, it increases storage by turning a plain wall into a viable resource for more space.

Kitchens

Above: Gleaming pots and pans can become decorative accessories when stored in full view. In this sparkling kitchen, a collection of stainless steel and copper cookware is displayed against an elegant marble-clad wall. Built-in storage bins, styled after the food bins found in general stores a hundred years ago, house such staples as salt, flour, and coffee.

Living and Dining Areas

Unlike the kitchen or the bedroom, the living room doesn't have the clear-cut function it once possessed. Coupled with the demise of the formal dining room and the rise of combination spaces that incorporate both dining and living areas, this entire part of the home is open to new interpretations. Living areas are used in more ways than ever before, which greatly influences the way they are furnished and the kinds of storage they include.

Living rooms are rarely the stiff showpieces they once were, since few of us have space to waste on a room that is used only when company comes. We truly live in these rooms these days, and so they often incorporate home offices, audiovisual equipment—which alone entails a huge stock of things that need to be stored—and reading materials. Plus there are the objects we want to have on show, such as prized possessions, artwork, and collectibles, which necessitate another kind of storage in the form of display.

Dining rooms have also changed, and now are rarely spaces devoted solely to one pursuit. In addition to living rooms, they can also be combined with libraries, entertainment areas, home offices, family rooms, and even kitchens, though this can be a difficult feat if a semblance of formal decorum is important to the homeowner. So, along with the usual trappings of a dining room—china, silver, table linens, serving pieces—there are now all sorts of other items that must fit into the space.

Carving out good-looking storage for all of these belongings can be challenging. It can be provided by built-ins or freestanding units, or an eclectic combination of both. Using two kinds of storage can add character and variety to a space. While all storage units are similar in substance, since they all have the same general purpose, they can be quite different in style. Built-ins can be fashioned in any decorative style from Colonial to Arts and Crafts and beyond. A considerable assortment of freestanding pieces is also available, whether you prefer the soft, weathered look of a vintage armoire or the sleek angularity of a postmodern entertainment unit.

Opposite: BUILT-IN SHELVING, ONE OF THE MOST EFFECTIVE TYPES OF STORAGE FOR ANY ROOM, CAN TAKE MANY FORMS. SOME OF THE MOST INTERESTING APPLICATIONS OF SHELVING ARE FOUND IN LIVING ROOMS, WHERE PEOPLE WANT TO MAKE THE BEST IMPRESSION. IN THIS COMFORTABLE ROOM, THE CUPBOARDS AND SHELVES SURROUNDING THIS FIREPLACE HAVE BEEN PAINTED AN ELEGANT PALE YELLOW, WHICH KEEPS THE STORAGE SYSTEM FROM SEEMING TOO INFORMAL.

Living and Dining Areas

Manipulating storage to suit your concept for the room's design is another issue that needs to be addressed. Both built-ins and freestanding pieces can be allowed to fade into the background, or they can be used to define a room. Natural recesses, such as alcoves, the spaces surrounding mantels, or whole blank walls, are ideal and unobtrusive spots for both built-in and freestanding storage, especially if the latter fits the niche exactly and mimics a built-in. Large shelving systems and shorter units placed back to back can also become room dividers, and these options have the added convenience of being accessible from both spaces.

Right: A storage system can become a source of display and a decorative device in its own right. In this dining room, a shallow gridlike system is used as an overlay to blanket a whole wall, turning an assortment of glassware and silver into a well-presented, carefully edited collection. The result is much more than the sum of its parts, greatly heightening the beauty and prestige of the pieces. The design also proves that clean, simple lines transcend period styling, for the amorphous grid is right at home in this ornate Georgian setting.

Living and Dining Areas

Below: Traditional Japanese tansu chests, which can be stacked in several configurations, have many applications in the home today, thanks to their clean lines and flexible design. In this dining room, several sections are pieced together to make a stunning but extremely functional storage system. The chests also provide display space for a collection of blue and white porcelain.

Above: A unique piece of furniture can do wonders for a room, harbor all sorts of items, and make the most of available space. In this case, the sizable Asian armoire is dramatic enough to add panache while offering up loads of storage, and a weathered wooden wine rack turns the wine bottles into a decorative statement.

Living and Dining Areas

Above: Built-ins and traditional furniture pieces aren't always necessary or even desirable. Instead of opting for a conventional matching dining room set in this space, an ensemble is pieced together with striking individual items, such as an interesting table and two different sets of chairs. In lieu of a breakfront or china cabinet, a matching pair of Asian-influenced armoires serves as storage and becomes the pièce de résistance of the room. The use of pairs (as in the chairs and armoires) also lends the room symmetry and balance.

Living and Dining Areas

36

Living and Dining Areas

Below: Think creatively about the storage potential of every type of container. While this antique cabinet was initially crafted for use as a dry bar, it can easily fit into a contemporary setup and be adapted to other tasks. For instance, a similar unit could be used to accommodate office supplies, compact disks, or anything that needs to be kept out of sight.

Opposite: A large storage center can be executed in any style. Here, a contemporary unit with high-tech capabilities is given an antique aura, thanks to its styling. When the doors are closed the cabinet resembles a handsome sideboard—no one would guess that it's an entertainment center.

Above: Armoires can be wonderful solutions for housing stereo equipment and collections of compact disks. They are available in a wide range of styles, from casual country to streamlined modern, which makes it easy to find one to suit your taste and decor. This handsome oak example has the clean lines and good looks of a traditional buffet, but it serves up music rather than food.

LIVING AND DINING AREAS

Left: It pays to break with convention. Generic folding metal chairs such as these would be an eyesore stacked against the wall, but when hung on the wall they make a stunning, graphic display. In fact, they turn the proverbial sow's ear into a silk purse. Best of all, when they are needed, they are a mere arm's length away.

LIVING AND DINING AREAS

Below: Built-in storage can be used to define activity areas as well as to accommodate all the essentials. In this case, the two functions go hand in hand. Open shelving at one end of the space holds an extensive library, while closed cabinets wrap around a snug corner to create a media center and bar area. A comfortable chair anchors the space.

Above: This elegant living room has two interesting storage solutions, one hidden and one in plain view. A small piece styled after a tansu chest is used as an end table next to the sofa to provide a home for odds and ends. An entire collection of art photographs is displayed on the wall rather than using the more common practice of hanging just a few; the black and white theme and simple frames keep the space from feeling cluttered.

LIVING AND DINING AREAS

Above: A LARGE PIECE SUCH AS THIS CAN PROVIDE LOTS OF STORAGE IN A KITCHEN OR DINING ROOM WITHOUT TAKING UP MUCH FLOOR SPACE. THE OPEN SHELVES HOUSE DISHES AND PRESERVES, WHILE A GLASS-FRONT CABINET IN THE CENTER KEEPS BAKED GOODS FRESH. THE DRAWERS CAN HOLD SILVERWARE AS WELL AS COOKING UTENSILS AND THE CUPBOARDS PROVIDE PLENTY OF SPACE FOR TABLE LINENS.

Below: A BIT OF COVER ALWAYS COMES IN HANDY WHEN STORAGE IS INVOLVED. HERE, A CLEVER SLIDING PANEL THAT HANGS FROM WHEELS RATHER THAN MOVING ON A TRACK CAN BE USED TO MASK ANY UNKEMPT COMPARTMENTS. THE WHOLE UNIT IS ALSO MOUNTED ON WHEELS, MAKING IT EVEN MORE FLEXIBLE AND MULTIFUNCTIONAL. IT CAN BE USED IN ANY SORT OF ROOM OR EVEN MOVED FROM PLACE TO PLACE AS NEEDED.

Living and Dining Areas

Right: Though not as flexible as the tansu chest, the quintessential armoire can also be used with other pieces to create a wall of storage. In this dining room, a tall, narrow version is paired with a squat cupboard against one wall to create a storage system that almost appears to be a set. Though these pieces don't match, they look good together because of their simple styling and similar hues.

LIVING AND DINING AREAS

Left: Bright color and good organization can make even the most basic storage systems seem exciting. Thanks to its cheerful blue, green, and yellow color scheme, this home office becomes a much more pleasant place to work.

LIVING AND DINING AREAS

Below: FIREWOOD NEED NOT BE AN EYESORE. AN INGENIOUS PERSON BUILT A HANDSOME WOODEN SHELF JUST A FEW FEET FROM THE FLOOR IN THIS ROOM TO PROVIDE A HOME FOR THE FIREWOOD AND A DISPLAY SPACE FOR A BEAUTIFULLY PAINTED WOODEN DUCK.

Above: SHELVES THAT ARE TOO SHALLOW TO PROVIDE MUCH STORAGE CAN BE DIVIDED INTO GRIDS TO BECOME WONDERFUL DISPLAY SPACES. HERE, PLAIN WHITE SHELVES HOLDING A COLLECTION OF PAINTINGS, CERAMICS, AND SHAKER BOXES CREATE THE APPEARANCE OF A GALLERY PRESENTATION.

Bedrooms and Baths

Space is always a precious commodity in a house, which may explain why bedrooms and bathrooms are gradually being redesigned to serve more than one purpose. While these rooms are still devoted mainly to the items required for sleeping, dressing, and bathing, they can also incorporate the furnishings, equipment, and accessories of many other pursuits. Shelves filled with books, cushy chairs for reading, and the fixings for home offices or gyms are often fitted into these private spaces in the home. In fact, we cram so many functions into bedrooms and bathrooms these days that their organizational perfection is imperative.

The one thing that all bedrooms have in common is, of course, the bed, which can—and often does—dominate the space. The way this furnishing is situated in the room dictates storage options. For instance, placing a bed against the center of a wall can create alcoves on either side for nightstands or dressers, while moving it toward one corner leaves room for larger storage units. Setting the bed in the middle of a larger room will give you lots of creative leeway, since a dressing room, private study, or home office can be created on any side of the bed. Both freestanding and built-in pieces can be integrated into all of these potential layouts. The area beneath the bed can become another source of storage; boxes and trunks can be hidden under a standard bed, while others can be constructed with drawers or cupboards in the space between the mattress and the floor.

Closets are also up for grabs these days—anything goes as long as it makes the most of a space. Many closets are outfitted with custom-made built-ins that mine every square inch of space. Others are sensibly arranged with economical wire systems that combine hanging fixtures and stacking baskets. These wire systems are just as effective as built-ins and easier to adapt to changing needs. Extra-large closets can even double as dressing rooms if there is enough space to move around comfortably. The only other requirements are a good mirror and adequate lighting.

Opposite: When storage is at a premium, it pays to take advantage of every option. Consider choosing a bed platform with drawers underneath instead of a box-spring bed set, and think about placing a small set of drawers beside the bed in lieu of a nightstand or table. In this case, both pieces match the wall behind the bed and add the warm glow of blond wood to the room.

Perhaps the biggest strides have been made where bathrooms are concerned, since many have more square footage than ever before. Often these generous spaces incorporate large sink areas or expansive vanities with ample storage underneath. Other large bathrooms have whole walls devoted to built-in storage in the form of cabinets and shelves. However, integrating storage into smaller bathrooms still takes ingenuity and planning.

With a little imagination, it's possible to "create" space where there doesn't seem to be any. Nooks between the ends or sides of tubs and the walls can be used for towel storage. Small chests or narrow but deep cabinets can be placed under sinks. Standard shelves can be built over toilets, and triangular shelving systems can fit into corners. Space is there for the taking in all of these rooms, with a little creativity and a lot of sensible design.

Left: OUT OF SIGHT, OUT OF MIND. ONCE THE DOORS ARE CLOSED ON THIS RICH MAHOGANY VANITY, WHICH WAS DEVISED TO MATCH THE REST OF THE WOODWORK IN THE ROOM, THE SPACE BECOMES A FORMAL SITTING AREA OFF A MASTER SUITE. THERE IS ALSO EXTRA STORAGE UNDER THE WINDOW SEAT.

Opposite: IN LIEU OF A SET WITH MATCHING COMPONENTS, EVERYTHING FROM THE BED TO DRESSERS CAN BE BUILT IN. IN THIS CASE, THE CHOICE OF BUILT-INS RATHER THAN FREESTANDING STORAGE ALSO MAKES THE MOST OF A RELATIVELY SMALL ROOM AND FURTHERS THE MINIMAL DESIGN AESTHETIC OF THE ARCHITECTURE.

BEDROOMS AND BATHS

Bedrooms and Baths

Opposite: Sometimes it pays to keep storage totally unobtrusive. In this bedroom, a whole wall of built-in cabinetry is so subtle it's almost indiscernible, which allows the pine plank paneling to take center stage and instills the room with an elegantly austere aesthetic. It also fits beautifully with the minimal architectural styling of the space.

Above: When space for storage and display seems nonexistent, put walls to work. In this small bathroom, a system of shelves built into an alcove provides plenty of open storage. Fluffy white towels are kept next to the shower, where they're needed most. A glass table takes the place of a medicine cabinet and allows attractive bottles of perfume to be on display.

Above: Too much of anything can be overwhelming. The careful balance of surfaces and elements keeps this entire expanse of built-in storage in a large bathroom in check. The bottom of the unit has a visually appealing and useful combination of cabinets and drawers, while the oversized mirror is complemented by the charming bank of glass-enclosed display cabinets.

Left: Make the most of every inch of closet space. In this closet, high ceilings provided ample opportunity to carve out lots of extra nooks and crannies. Shelves top double-hung clothes racks and a high chest, while the extra-tall closet doors were also mined as a source of storage and now house dozens of pairs of shoes.

Right: Don't hide an attractive tableau behind closed doors. A sumptuous storage cabinet filled with items that are visually appealing should be left open for all to see. Here, an equally attractive antique settee is used to prop the door in place on a permanent basis.

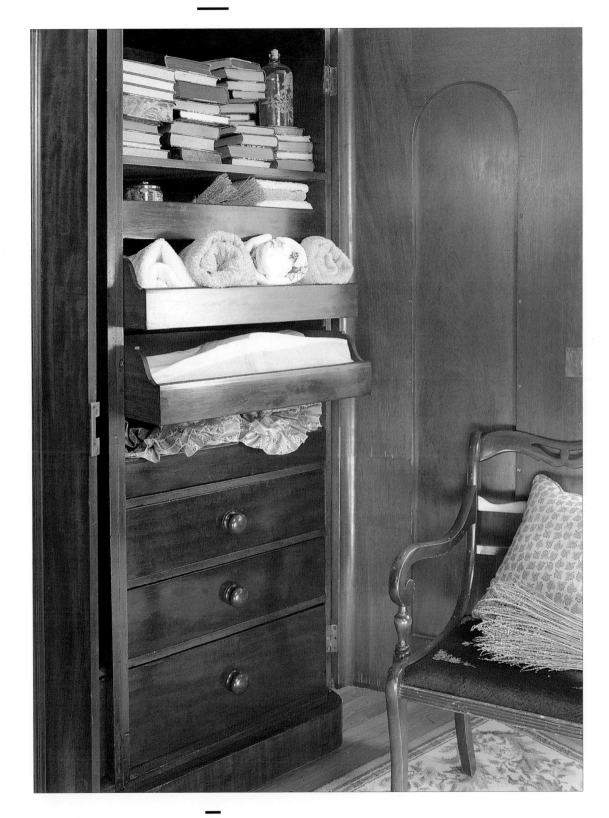

Bedrooms and Baths

Opposite: Storage can get gussied up, as evidenced by this child's desk system fitted out to resemble an apartment building. The shelving unit that splits the desk into two is both well planned and broadly appealing, since it is accessible from either work station and neutral enough to appeal to a wide age range.

Below: Collectibles and interesting objects can lend a room depth and character, so make sure they show. In this cozy bedroom, an assortment of mismatched shelving units is coupled with some creatively recycled storage options, such as a rustic basket, an antique chest, and a vintage suitcase. The quirky combination yields charming results.

Above: A child's room is the perfect place to blend storage with a touch of whimsy. This is done with a variety of different shelving options, all of which serve to keep items out of the way yet still within reach. In addition to their functional mission, these shelves are a thoughtful decorative device, since they highlight the interesting items and forge them into an engaging tableau.

Bedrooms and Baths

Left: Towels should be stored right where they're needed rather than out in the hallway in a linen closet. In this case, the decorative aspect of the bathroom is also furthered by the device: glass shelves keep the alcove next to the sink open and airy.

Below: A sleek, minimal system devised to fit under this double sink offers up adequate storage and makes innovative use of dead space. It combines a single set of deep, capacious drawers with steel rails that can be used as towel bars if necessary. It is complemented by another steel towel rack over the tub of the sort that is usually reserved for hotels but is an ingenious addition in this modern bath.

Opposite: A number of exotic elements with wildly divergent styles are combined in this eclectic bathroom, from a glass block wall to a stepped storage unit styled after a tansu chest. The room is anchored with a deceptively simple counter sheathed with neutral marble tiles and fitted with contrasting drawers for storage.

Above: Small closets can hold a tremendous amount if they're well planned and organized. Here, two bars have been installed to hold short items such as jackets, rather than the traditional single, high bar that holds longer items but fewer of them. A built-in unit has cubbyholes for shoes, caps, and videotapes in addition to shelves and drawers.

Below: When built-in cabinets or cupboards aren't an option, attractive freestanding pieces can provide the necessary storage. Here, a small trunk made of reeds holds an assortment of sponges and washcloths, while a vintage basket keeps towels handy.

Bedrooms and Baths

Above: Necessary incidentals can be forged into attractive displays and stored where they do the most good. An assortment of colorful soaps makes a pretty display on a table next to a sink in a powder room. The table provides a little hidden storage too, in the form of a single drawer that can hold additional toiletries.

Special Spaces

Planning storage systems for specific rooms is only the first step to maximizing your home's storage potential. There are many other parts of a house that can be organized or adapted in the same way for optimum storage. Try to make the most of every existing area, be it an attic or basement, a storage locker in an apartment or condominium, an extra-large hallway or landing, a niche under the stairs or eaves, a mud or utility room, or an oversized garage or shed.

Attics and basements can add significant living and storage space to a home, and often these areas incorporate both. Whether you set up a home office or play room in an attic or a craft corner and hobby center in a basement, you will need room for the designated activities and storage for the accoutrements that go with them. Often the storage component of the space can be used to define activity centers. Shelving units can become freestanding walls that separate areas, which allows each to have its own personality and function. Attics and basements also offer ideal spots for seasonal storage, and again, thoughtful design and good planning can make these spaces useful for other purposes as well.

Extra-large hallways and landings, as well as the triangular alcoves under stairs or eaves, are "found" spaces. They're often expansive enough to double as storage areas, freeing up floor space in living areas. Perhaps an entire library can fit along a wall that forms one side of a wide staircase, or it might line a sizable second-floor landing. A small home office, a closet, or several cupboards might be worked into the oddly shaped spaces under stairs. Because these nooks and crannies are actually alcoves and recesses, they can accommodate built-in or set-in storage systems fairly unobtrusively.

Mud and utility rooms, along with oversized garages and storage sheds, are often neglected spaces that just beg for attention. Instead of letting them become cluttered

Opposite: Landings can be turned into wonderful display spaces to allow you to store some of your favorite items in plain sight. In this stairwell, a ledge topped with rich wood becomes a perfect spot to show off a collection of pottery vases. The skylight above the ledge provides plenty of light for this little "gallery."

zones or oversized junk drawers, use them to their fullest potential. Mud and utility rooms are attached to residences rather than constructed as outbuildings, so they are naturals to mine as an extra source of convenient storage for frequently used items. Garages and sheds are perfect areas in which to keep seasonal belongings or to set up craft centers, especially if the house itself is short on space. Like the rest of the home, these special spaces can and should be outfitted with attractive and carefully planned built-in or freestanding storage. It is more sensible, however, to opt for cost-effective systems, since most likely these areas will not be on view.

Left: A HALLWAY BECOMES A SIGNIFICANT—AND STUNNING—SOURCE OF STORAGE WHEN FITTED WITH CUSTOM-DESIGNED CABINETRY. THE BUILT-INS RUN THE ENTIRE LENGTH OF THE SPACE AND ALMOST FROM FLOOR TO CEILING, BUT THEY STOP JUST SHORT OF THE BEAMS TO SHOW OFF THE DRAMATIC ANGLED CEILING. THE STREAMLINED SHUTTER-STYLE DOORS BECOME A DECORATIVE ELEMENT THAT GIVES THE SPACE AN ELEGANT DEMEANOR.

Special Spaces

Right: A good storage system can organize and unclutter a laundry room, making it attractive and utilitarian at the same time. The shelves in this laundry room are placed where they can do the most good: over the granite counter that tops the washer and dryer. To keep the rest of the space equally tidy and appealing, an open metal shelving unit was turned into a "closed" system by fitting the shelves with baskets in varying styles and sizes.

Special Spaces

Left: An artist's studio should have shallow, outsized drawers that can store works of art, and the stairway in this space offers a perfect place to incorporate that sort of storage. More importantly, it allows a small corner of a room to become a fully functional studio, since supplies can be stored right next to the easel, where they're needed most.

Special Spaces

Above: This stylized shelving system owes its sheer utility to the architecture of the space, which makes it easily accessible from the stairs. Its sophisticated appeal is due to the clever blend of materials used for the shelving: thick, white laminate, heavy plate glass, and thin stainless steel supports. The shelves are actually quite shallow, but they incorporate plenty of spaces to display a collection of ceramics and a small library of art books.

Below: Make the most of unusual spaces with a little ingenuity. An idiosyncratic alcove at the top of a narrow staircase would be considered dead space by most homeowners, but there are storage opportunities to salvage here. Floor-to-ceiling shelves transform the nook into a wonderful display spot for a collection of vintage boxes and tins.

Special Spaces

Above: A long wall that could be a design deficit is turned into a narrow but spacious home office for two. The large double desk has plenty of drawers for office equipment and files, while the larger cabinet in the middle acts as both a storage unit and a display space.

SPECIAL SPACES

Right: With proper furniture placement and the right decorative devices, a small room can be turned into a complete home office without seeming overcrowded. The office furnishings in this room are large, well designed, and offer plenty of storage. A simple grouping of plump armchairs in rich russet hues and bold stripes keeps these furnishings from overwhelming the space, and the whole arrangement is set off and defined by an equally bold and colorful rug.

Special Spaces

Below: An attic landing can become more than a mere pass-through, especially if it is fairly large or has ample wall space. Built-in bookshelves can often be fit into such a place. In this case, the landing was sizable enough to house an expansive set of shelves. The unit reaches floor to ceiling, but the dramatic slant of the roof makes for a charming variety of shelf heights.

Above: Spaces under the eaves often provide interesting and abundant opportunities for storage, as is evident in this home office that combines several types of storage. Built-in shelves house books and knickknacks, while a built-in banquette has room for baskets and bins underneath. The area with the highest ceiling holds a full-sized desk, which is also packed with storage potential.

Special Spaces

Above: Attics are certainly ideal storage areas, but they can also provide space for a hobby center when the storage units are well organized. Vintage trunks are an attractive alternative to cardboard boxes or plastic bins. In this attic, they've been stashed in the areas where it is impossible to walk upright, leaving the most functional part of the room for other pursuits.

Special Spaces

SPECIAL SPACES

Below: Utility rooms often become the repositories of unsightly cardboard boxes filled with unwanted clothes and appliances, but this doesn't have to be the case. With a little ruthless paring down and a bit of organization, the space can become the perfect hobby area. This utility room is quite small, but the addition of several shelves, a deep sink, and an assortment of baskets has turned it into a charming "shed" for an avid gardener.

Opposite: A narrow breezeway can't accommodate a closet, but it can be outfitted with a pegboard and a bench to afford it all the easy-access storage needed. In addition to providing a place to sit, the bench brings order to all the footwear. It's also a handy spot for the occasional backpack or piece of sporting equipment.

Above: Given their sheer volume, hats have the potential to be a space drain. Hanging them on a pegboard over a closet door turns them into an attractive display and keeps them within reach. Best of all, they can be crowded together with no harm done.

Sources

ARCHITECTS AND INTERIOR DESIGNERS

(page 2)
Marmol and Radziner
  Architects
Los Angeles, CA
(310) 364-1814

(page 6, 66 right)
Anne Lenox
Partners In Design
Newton Center, MA
(617) 969-3626

(page 10)
Weston & Hewitson
  Architects
Hingham, MA
(781) 749-8587

(pages 14, 15)
Ted Wengren
(207) 865-1470

(page 17 right)
Kitchen Design Studio
New Canaan, CT
(203) 966-0355

(pages 18, 19)
Michaela Scherrer Interiors
Pasadena, CA
(818) 240-2315

(page 23)
Mark Mack Architects
Santa Monica, CA
(310) 822-0094

(page 26 left)
Chip Dewing
Dewing & Schmidt Architects
Cambridge, MA
(617) 876-0066

(page 28)
Diana Sawichi Interior
  Design, Inc.
Westport, CT
(203) 454-5890

(page 35)
McDonald & Moore, Ltd.
San Jose, CA
(408) 292-6997

(page 37 right)
Barbara Goldfarb
Design Logic
Red Bank, NJ
(732) 842-2922

(page 39 bottom)
Paul Dandini, Architect
P.M. Dandini Construction
  Company Inc.
Somerville, MA
(617) 623-8888

(page 44)
Barbara Barry
Los Angeles, CA
(310) 276-9977

(page 55 right)
Lloyd Jafvert, AIA
Bloomington, MN
(612) 897-5001

(page 58)
Steven Tucker
Boston, MA
(617) 338-7300

(page 60)
Deborah T. Lipner, Inc.
Greenwich, CT
(203) 629-2626

(page 64)
Marc Melvin
San Francisco, CA
(415) 567-1484

(page 65)
Agnes Bourne
San Francisco, CA
(415) 626-6883

(page 46)
Carole Kaplan
Two By Two Interior Design
Andover, MA
(978) 470-3131

(page 49 right)
Peter Labau
Classic Restorations
Watertown, MA
(617) 923-0505

(page 55 left)
Catalano Architects
Boston, MA
(617) 338-7447

(page 61)
Elliott & Elliott Architecture
Blue Hill, ME
(207) 374-2566

(page 63 left)
Lloy Hack Interior Design
Boston, MA
(617) 247-8773

Sources

(page 68)
Toth Design
Concord, MA
(978) 369-3797

Suppliers

The phone numbers listed can be called for store locations and/or catalogs.

California Closets
(800) 225-6901
www.calclosets.com

The Container Store
(800) 733-3532
www.containerstore.com

Crate & Barrel
(800) 323-5461

Hold Everything
(800) 421-2264

Ikea
(800) 434-4532
www.ikea.com

Levenger
(800) 544-0880
www.levenger.com

Spiegel
(800) 345-4500
www.spiegel.com

Photography Credits

B&B Italia: p. 30 (designer: Antonio Citterio)

©Phillip Ennis: pp. 37 right (designer: Barbara Goldfarb, Design Logic), 50 (designer: George Constant)

©Michael Garland: pp. 18 (designer: Michaela Scherrer), 19 (designer: Michaela Scherrer), 31 right (designer: Carla Schrad), 34 left (designer: A.J. Killawea), 43 right (designer: Alice Fellows), 52 left, 69 right (designers: Kip and Sherna Steward)

©Tria Giovan: pp. 26 right, 32, 51, 56 left, 56 right, 57

©David Henderson: pp. 21 left, 36

©Nancy Hill: pp. 17 right (designer: Kitchen Design Studio), 21 right, 28 (designer: Diana Sawichi Interior Design, Inc.), 54 (architect: Robert Nevins), 55 right (architect: Lloyd Jafvert), 60 (designer: Deborah T. Lipner, Inc.)

©www.davidduncanlivingston.com: pp. 25 right, 31 left, 35 (designers: McDonald & Moore), 64 (designer: Marc Melvin), 65 (designer: Agnes Bourne)

©Eric Roth: pp. 6 (designer: Anne Lenox, Partners In Design), 24, 25 left, 26 left (architect: Chip Dewing, Dewing & Schmidt Architects), 37 left (courtesy Circle Furniture), 39 bottom (architect: Paul Dandini), 46 (designer: Carole Kaplan, Two By Two Interior Design), 49 right (designer: Peter Labau, Classic Restorations), 55 left (architect: Catalano Architects), 63 left (designer: Lloy Hack Interior Design), 66 right, 68 (designer: Toth Design)

Spiegel: pp. 39 top, 40 right

©Tim Street-Porter: pp. 2 (architects: Marmol and Radziner), 23 (architect: Mark Mack Architects), 44 (designer: Barbara Barry), 47 (architect: Josh Schweitzer), 53

©Brian Vanden Brink: pp. 10 (architects: Weston & Hewitson), 13 (designer: Lou Ekus), 14–15 (architect: Ted Wengren), 16 (designer: Karin Thomas), 58 (architect: Steven Tucker), 61 (architects: Elliott & Elliott), 67

Elizabeth Whiting Associates: pp. 9, 12, 13, 15 right, 17 left, 20, 22, 27, 33, 34 right, 38, 40 left, 41, 42, 43 left, 48, 49 left, 52 right, 62, 63 right, 66 left, 69 left

Index

Alcoves, 59
  shelves built into, 49, 63
Armoires, 37, 41
  Asian, 34, 35
Artist's studio, 62
Attics, 59, 66, 67

Banquette, 66
Bar area, 37, 39
Basements, 59
Baskets, 12, 14, 25, 56, 69
Bathrooms, 45, 46
  concealed storage in, 49, 54–55
  freestanding storage in, 56, 57
  open storage in, 49, 55–57
Bed
  placement of, 45
  platform, with drawers, 44
Bedrooms, 45
  concealed storage in, 44, 46–48
  freestanding storage in, 51
  open storage in, 51, 52
Bookshelves, 30, 31, 66
Built-in cabinetry
  in bathroom, 49, 54–55
  in bedroom, 44, 47, 48
  in hallway, 60
  in kitchen, 11, 16–19, 23–24, 27
  in living room, 28, 29, 39

Cabinet(s)
  in bathroom, 49, 54–55
  in bedroom, 44, 47, 48
  glass-fronted, 10, 19, 20
  in hallway, 60
  in kitchen, 10, 11, 14, 16–19, 23–24
  medicine, 21
  in office, 9, 64, 65
Canisters, glass, 18, 20
Chairs, hung on wall, 38
Children
  desk system for, 53
  play room for, 52
  storage needs, 7–8
Closets, 45, 50, 56
Concealed storage, 8
  in bathroom, 49, 54–55
  in bedroom, 44, 46–48

  in dining room, 40, 41
  in kitchen, 10, 11, 14, 16–19, 23–24
  in living room, 28, 30, 36–37, 39
  in office, 9, 64–65
Cupboards
  built-in, 17, 28
  vintage, 25

Desk systems
  child's, 53
  for two, 64
Dining rooms, 29
  concealed storage in, 40, 41
  freestanding storage in, 35, 40
  open storage in, 32, 38
Drawers
  bathroom, 54
  under bed, 44
  under stairway, 62

Electronics, storage for, 30, 36, 37, 39

Fireplaces
  cupboards surrounding, 28
  nonworking, 32
Firewood, storage for, 43
Freestanding storage
  armoires, 34, 35, 37, 41
  in bathroom, 54, 56, 57
  in bedroom, 51
  in dining room, 35, 40
  in kitchen, 11–12, 25
  in living room, 29, 36–37, 39
  media centers, 30, 36, 37
  tansu chests, 6, 34, 39, 54
  on wheels, 40

Garages, 59–60
Gridlike shelving, 33

Hallways, 59
  with built-in cabinets, 60
  freestanding storage in, 6
  open storage in, 68, 69
Hobby centers
  in attic, 67
  in utility room, 69

Home offices, 64–66
  metal cabinet in, 9
  shelves in, 42
Hooks, kitchen, 12, 26

Islands, kitchen, 10, 14, 16, 24

Japanese storage chests, 6, 34
  pieces styled after, 39, 54

Kitchens, 10, 11–12, 12–27
  concealed storage in, 10, 11, 14, 16–19, 23–24
  open storage in, 12–15, 18, 22–23, 26–27

Landings, 59
  with bookshelves, 66
  as display spaces, 58
Laundry room, 61
Living rooms, 29
  concealed storage in, 28, 30, 36, 37, 39
  open storage in, 28, 30–32, 39

Media centers, 30, 36, 37
Medicine cabinet, 21
Metal cabinet, 9

Offices, See Home offices
Opaque storage, 30
Open storage. See also Shelves/shelving
  in bathroom, 49, 55–57
  in bedroom, 51, 52
  in dining room, 33, 38
  display value of, 8
  in hallway, 68, 69
  in kitchen, 12–15, 18, 22–23, 26–27
  in laundry room, 61
  in living room, 28, 30–32, 39
  in office, 66

Pegboards, 68, 69

Racks
  in bathroom, 55
  in kitchen, 12, 21, 26

Room dividers, 30

Sheds, 59–60, 69
Shelves/shelving
  in bathroom, 49, 55
  in bedroom, 52
  bookshelves, 30, 31, 66
  bracket system, 22
  in child's room, 52
  circular, 17
  in dining room, 33
  glass and steel, 18
  gridlike, 33
  in home office, 42, 66
  industrial, 12
  in kitchen, 10, 12–15, 18, 22–23, 26–27
  in laundry room, 61
  in living room, 28, 30–31, 39, 40
  with moveable hooks, 26
  sculptural, 23, 63
  shallow, 43, 63
  with sliding panel, 40
  along staircase, 63
  tension suspension, 31
Shoes, storage for, 50, 56
Sinks
  bar, 21
  bathroom, 54–55
Space planning, 8–9
  kitchen, 11
Staircase
  drawers under, 62
  shelving, 63
Suitcases, vintage, 52, 67

Tansu chests, 6, 34, 39, 54

Utility rooms, 59–60, 69

Vanity, 46

Window seat, storage under, 46, 66
Wine racks, 25, 34